# RISK MANAGEMENT PROGRAM GUIDANCE FOR
# PROPANE STORAGE FACILITIES
# (40 CFR PART 68)

This document provides guidance to owners and operators of stationary sources to determine if their processes are subject to regulation under section 112(r) of the Clean Air Act and 40 CFR part 68 and to comply with regulations. This document does not substitute for EPA's regulations, nor is it a regulation itself. Thus, it cannot impose legally binding requirements on EPA, states, or the regulated community, and may not apply to a particular situation based upon circumstances. This guidance does not represent final agency action, and EPA may change it in the future, as appropriate.

# TABLE OF CONTENTS

# TABLE OF POTENTIALLY REGULATED ENTITIES

*This table is not intended to be exhaustive, but rather provides a guide for readers regarding entities likely to be regulated under 40 CFR part 68. This table lists the types of entities that EPA is now aware could potentially be regulated by this rule and covered by this document. Other types of entities not listed in this table could also be affected. To determine whether your facility is covered by the risk management program rules in part 68, you should carefully examine the applicability criteria discussed in Chapter 1 of this guidance and in 40 CFR 68.10. If you have questions regarding the applicability of this rule to a particular entity, call the EPCRA/CAA Hotline at (800) 424-9346 (TDD: (800) 553-7672).*

| Category | NAICS Codes | SIC Codes | Examples of Potentially Regulated Entities |
|---|---|---|---|
| Propane manufacturers and processors | 32411 32511 | 2911 2865 2869 | Petroleum refineries Petrochemical Manufacturers |
| Propane wholesalers | 42271 42272 | 5171 5172 | Bulk stations and terminals Other petroleum product wholesalers |

# INTRODUCTION

This guidance is intended for propane storage facilities, such as wholesale distribution facilities and bulk storage terminals. This document is a revision of EPA's October 1998 guidance for propane storage facilities. The document has been revised to reflect changes resulting from the 1999 Chemical Safety Information, Site Security and Fuels Regulatory Relief Act.

If you have more than 10,000 pounds of propane stored in a single vessel or in a group of vessels (tanks, cylinders) that are connected or stored close together, you may need to comply with the Chemical Accident Prevention rule issued by the U.S. Environmental Protection Agency (EPA) under the Clean Air Act. The rule is codified as part 68 of Title 40 of the Code of Federal Regulations (CFR). The goal of this rule is to prevent accidental releases that could affect the public or the environment. If you are subject to part 68 for propane storage, you must be in compliance no later than January 5, 2000, or the date on which you first have more than a threshold quantity of a regulated substance in a process, whichever is later.

If you have more than 10,000 pounds of propane, you are subject to part 68 unless one of the following applies to you:

- • The propane is stored for use as a fuel at your facility.
- • The propane is held for sale, and the facility is a retail facility. A retail facility is one at which more than half of the income is obtained from direct sales to end users or at which more than half the fuel sold, by volume, is sold through a cylinder exchange program.

If you have more than 10,000 pounds of propane and you manufacture propane, use it as a feedstock, or store it in bulk for sale, other than to end users, or if your retail sales do not make up more than half of your income, you are subject to part 68. If you manufacture or use propane as a feedstock or store it for wholesale distribution and use it as a fuel, the propane used for fuel is not covered by part 68; the propane manufactured, processed, or stored for wholesale distribution is subject to part 68 provided the quantity is greater than 10,000 pounds. If you use propane to fuel a covered process containing other regulated substances above their thresholds, the propane is not covered, but you will have to consider the hazards created by the propane when you conduct your process hazard analysis or hazard review.

For most propane storage facilities, complying with this rule will be easy because most of the requirements are similar to those you already comply with under state or local rules based on the National Fire Protection Association (NFPA) standard number 58 on propane. If you are complying with NFPA-58 and implementing other safe engineering practices for propane, you should have little more to do for this rule besides filing a report with EPA.

## AM I COVERED?

The capacity of propane tanks is usually given as water capacity (this information should be on the nameplate of the tank). Table 1 translates the water capacity (in gallons) into propane weight (in pounds). Read the tank capacity on the nameplate and check this table.

Table 1 assumes that tanks are filled to 88 percent of capacity, the maximum level allowable under NFPA-58 at 60• F. If you always keep your tanks filled to a lower level, you should adjust these numbers to reflect your lower inventory. If you have larger tanks, multiply the water capacity times 3.696 to determine pounds at 88 percent capacity.

### TABLE 1
### TYPICAL WATER CAPACITY OF PROPANE TANKS, TRANSPORTS AND RAILROAD TANK CARS

| | Capacity in Gallons | Capacity in Pounds |
|---|---|---|
| Storage Tank | 12,000 | 44,400 |
| Storage Tank | 18,000 | 66,500 |
| Storage Tank | 30,000 | 111,000 |
| Storage Tank | 40,000 | 148,000 |
| Storage Tank | 60,000 | 222,000 |
| Storage Tank | 100,000 | 370,000 |
| Storage Tank | 120,000 | 444,000 |
| Transport (Cargo Tank) | 8,000 - 11,000 | 30,000-41,000 |
| Rail Car DOT Class 105J300W & 105A300W | 11,000 - 34,500 | 41,000-128,000 |
| Rail Car DOT Class 112J340W & 112T340W 114J340W & 114T340W 114J400W & 114T400W | 26,000 - 34,500 | 96,000-128,000 |
| Bobtails (Delivery Cargo Tank) | 750 - 3200 | 2,800-12,000 |

Also, add up the amounts of propane in tanks that are connected or close together. If you have four 750-gallon tanks, you are subject to the rule if the tanks are connected by piping, or if you store the tanks close enough together that they could be involved in a single accident. For example, if a fire could spread to all the tanks, they are considered one "process," and the propane in all the tanks must be counted toward the 10,000-pound threshold. You should also consider whether, if one tank exploded, the other tanks could be affected by the shrapnel or fire. If you have several groups of propane tanks, widely separated, you may be considered to have multiple processes (see Appendix A for additional guidance on determining whether your propane tanks are separated far enough to be considered multiple processes.)

## WHAT DO I HAVE TO DO?

The first step you should take after determining that you are covered by the rule is to decide which Program level you are in. EPA developed the rule with three Program levels to reflect different levels of risk and levels of effort needed to prevent accidents.

- • **Program 1** is a minimal set of requirements for processes that have a very low risk of affecting the public in the event of an accident.
- • **Program 2** is a streamlined set of requirements for processes not eligible for Program 1 or subject to Program 3.
- • **Program 3** applies to processes that are not eligible for Program 1 and that are either subject to the Process Safety Management (PSM) Standard of the Occupational Safety and Health Administration (OSHA) or in certain industrial sectors (some chemical manufacturers, all refineries, and all pulp mills).

This document does not provide detailed guidance on Program 3. Propane storage facilities subject to part 68 will generally either be eligible for Program 1 or subject to Program 3. The OSHA PSM standard exempts all retail facilities, using the same definition of retail facilities given above, and fuel users, but covers wholesalers and manufacturers that process propane. Consequently, a propane facility subject to part 68 will generally also be subject to the OSHA PSM standard. If you are subject to the OSHA PSM standard, you should also see EPA's *General Guidance for Risk Management Programs* or OSHA's *Process Safety Management Guidelines for Compliance* (OSHA 3133).

If you can qualify a process for Program 1, it is in your best interests to do so, even if the process is already subject to OSHA PSM. For Program 1 processes, the implementing agency will enforce only the minimal Program 1 requirements. If you assign a process to Program 3 when it might qualify for Program 1, the implementing agency will enforce all of the requirements of the higher program level. If, however, you are already in compliance with the prevention elements of Program 3, you may want to use your RMP to inform the community of your prevention efforts.

4

# PROGRAM 1

## ELIGIBILITY

Some propane storage facilities will be eligible for Program 1, particularly those that are a considerable distance from any other business or residence. For a process to be eligible for Program 1, it must meet the following criteria:

- • The process must not have had an accidental release of propane that led to deaths or injuries of people offsite or response or restoration activities at environmental receptors in the last five years. Environmental receptors are limited to national or state parks, forests, or monuments; officially designated wildlife sanctuaries, preserves, refuges, or areas; and Federal wilderness areas; and,

- • There are no public receptors within a distance to a 1 psi overpressure from a worst-case release.

A worst-case release is defined by the rule as the loss of the contents of the single largest vessel (or piping) containing the regulated substance. For propane and other flammable substances, the released substance is assumed to explode and generate a pressure wave that can damage people or structures. The rule requires you to determine the distance to a 1 psi overpressure (at 1 psi, windows will break). This scenario is required by the regulation, and you must adopt this scenario. Table 2 provides the worst-case distance to a 1 psi overpressure for propane tanks.

**TABLE 2**
**DISTANCE TO A 1 PSI OVERPRESSURE**

| Nominal Water Capacity (Gallons) | Distance to Endpoint (Miles) |
|---|---|
| 500 - 1,750 | 0.1 |
| 1,751 - 7,000 | 0.2 |
| 7,001 - 23,000 | 0.3 |
| 23,001 - 51,000 | 0.4 |
| 51,001 - 90,000 | 0.5 |
| 90,001 - 120,000 | 0.6 |

If you have different quantities, RMP*Comp, a software system developed by EPA and NOAA, will allow you to calculate worst-case distances quickly. RMP*Comp can be downloaded from www.epa.gov/emergencies/rmp. You may also use other models to develop distance estimates (see 40 CFR 68.22 for the parameters you must use in estimating worst-case release distances). Next, you must determine if there are "public receptors" within a circle whose radius is equal to this distance. Public receptors include "offsite residences, institutions (e.g., schools and hospitals), industrial, commercial, and office buildings, parks, or recreational areas inhabited or occupied by the public at any time without restriction by the stationary source where members of the public could be exposed to toxic concentrations, radiant heat, or overpressure, as a result of an accidental release." Offsite means areas beyond your property boundary and "areas within the property boundary to which the public has routine and unrestricted access during or outside business hours." Public roads are not public receptors.

If there are no public receptors within the distance to a 1 psi overpressure for your largest vessel and the process has not had an accidental release that caused any of the listed offsite impacts, your process is eligible for Program 1. If you have questions about whether certain areas are considered public receptors, call the Emergency Planning and Community Right-to-Know Act (EPCRA) hotline at (800) 424-9346 (for DC area (703) 412-9810; T.D. (800) 553-7672) or check EPA's General Guidance for Risk Management Programs (available from the hotline or electronically at www.epa.gov/emergencies/rmp).

## WHAT MUST I DO FOR A PROGRAM 1 PROCESS?

Because your worst-case release would not affect public receptors, you only need to do two things:

✦ Coordinate emergency response with your local fire department and any other local emergency planning and response agencies; and,

✦ Complete a brief Risk Management Plan (RMP), as described below

Coordination with the fire department may consist of a discussion with them or a walk-through of your facility. The purpose is simply to be sure that the fire department is aware of the hazards associated with propane at your facility and ready to respond if an accident occurs. Also, contact your State Emergency Response Commission (SERC) to identify your Local Emergency Planning Committee (LEPC). You can get contact information for your SERC from the EPCRA hotline (noted above).

The RMP will be filed with EPA and made available to state and local agencies and the public. EPA has developed a web-based submission system, called RMP*eSubmit, that will make filing a RMP or making changes or corrections to an existing plan easy. You can access this information at www.epa.gov/emergencies/rmp. EPA's previous system for electronic submission - RMP*Submit - will no longer be available on the EPA website after March 2009.

The RMP includes registration information (basic facility information); the worst-case release scenario; a five-year accident history (of accidents that caused deaths, injuries, or significant property damage on site, known offsite deaths or injuries, offsite property or environmental damage, or evacuations or shelterings in place); emergency response activities; a brief executive summary; and a certification statement.

The executive summary should be a brief description of the facility, the worst-case release scenario, steps you take to prevent accidents (e.g., complying with state and local laws), emergency response information (e.g., your coordination with the fire department), and any steps you are planning to take to improve safety (e.g., upgrading equipment to meet newer editions of NFPA-58). The rest of the RMP is filling in names, addresses, and numbers, and checking appropriate boxes. You do not need to submit supporting documentation; you need only keep it onsite for inspection. Most propane storage facilities will not have any accidents to report on the five-year accident history. If you do not, you need not complete that section. A sample RMP for a Program 1 propane facility is attached.

# PROGRAM 3

If your process is not eligible for Program 1, the process is most likely in Program 3 because it is probably subject to the OSHA PSM standard. Most propane storage facilities that are in commercial or industrial areas or close to residential areas will be subject to Program 3.

## WHAT MUST I DO FOR PROGRAM 3?

For Program 3, you must:

- • Analyze both a worst-case release scenario and an alternative release scenario;
- • Implement a prevention program;
- • Implement an emergency response program if your employees will respond to a release; and
- • File an RMP.

## WHAT ARE THE RELEASE SCENARIOS?

**Worst-Case Scenario.** Part 68 defines the worst-case release scenario you must analyze. It is described in the previous section of this guidance (regarding Program 1). You can simply use Table 2 or RMP*Comp to define the distance to the 1 psi endpoint for your largest tank.

**Alternative Release Scenario.** An alternative release scenario is a scenario that is more likely to happen. It must reach an endpoint offsite unless no such scenario exists. One of the following scenarios may be appropriate for you.

- • **Pull-Away Explosion.** An alternative scenario may be a hose rupture caused by a pull-away. A pull-away can occur if the driver fails to remove the hoses between the storage tank and the transfer vehicle before moving the vehicle. In this scenario, the failure involves a 25-foot length of unloading hose, 4" in diameter. The active mitigation devices are assumed to work as designed, limiting the release to the contents of the hose. The release leads to a vapor cloud explosion (endpoint 1 psi). The quantity released is 69 pounds. The distance to the endpoint is 175 feet (report as 0.03 miles).

- • **Piping Break.** Another alternative scenario you may want to consider is a break in propane piping leading to a 10-minute release and explosion. The distance to the 1 psi endpoint is shown in Table 3.

Appendix B includes release calculations for alternative scenarios as well as scenarios in addition to the two provided above.

**TABLE 3**
**DISTANCES TO 1 PSI FOR PIPE RELEASES**

| Pipe Size (inches) | Quantity Released (pounds) | Distance to 1 psi |
|:---:|:---:|:---:|
| 0.5 | 4,738 | 0.1 |
| 1 | 18,951 | 0.2 |
| 2 | 75,804 | 0.3 |

You must estimate in the RMP residential populations within the circles defined by the endpoints for your worst-case and alternative release scenarios (i.e., the center of the circle is the point of release and the radius is the distance to the endpoint). You may use Census data and round to two significant digits (e.g., 1147 becomes 1100, and 123 becomes 120). You do not need to conduct surveys or correct Census data. In addition, you must report in the RMP whether certain types of public receptors (e.g., schools, hospitals) and environmental receptors are within the circle. You do not need to identify specific receptors; you simply need to check off the category.

## WHAT DO I HAVE TO DO FOR THE PREVENTION PROGRAM?

The Program 3 prevention program is identical to the OSHA PSM Standard and has 12 elements:

+ Safety information
+ Process Hazard Analysis
+ Operating procedures
+ Training
+ Mechanical Integrity (Maintenance)
+ Pre-startup reviews
+ Management of change
+ Compliance audits
+ Incident investigation
+ Employee participation
+ Hot work permit
+ Contractors

If you are complying with NFPA-58 or state or local laws based on it, following the guidelines in the National Propane Gas Association (NPGA) LP Gas Safety Handbook, and implementing NPGA safety bulletins, you are probably already doing most of what you need to do to comply with these requirements. The following sections provide additional information on how your current practices will help you comply with the EPA rule. For a more complete discussion of the requirements, see the *General Guidance for Risk Management Programs,* available at www.epa.gov/emergencies/rmp, and OSHA's *Process Safety Management Guidelines for Compliance* (OSHA 3133).

**Safety Information.** You must have up-to-date information on propane and your propane equipment. You must have a Material Safety Data Sheet (MSDS) on propane. If you do not have one, contact your supplier for a copy. You must also document your maximum intended inventory for your propane equipment. This will generally be the capacity of your tank or tanks; see Table 1.

You need information on safe upper and lower temperatures, pressures, flows, and compositions. The following information should meet this requirement:

- • Propane is a gas at normal temperatures and pressures. It is commonly liquefied by storing it in a closed container at pressures higher than its equilibrium vapor pressure. There is a direct relationship between ambient temperature and the pressure inside the storage container. As the ambient temperature increases, the pressure of the container increases proportionately. According to NFPA 58, 1998 Edition, Table B-1.2.1, commercial propane when heated to a temperature of 105•F will produce a pressure of 210 pounds per square inch, gauge (psig). NFPA 58, 1998 Edition, Table 2-2.2.2 sets the current minimum design pressure for an ASME tank at 250 pounds per square inch, absolute (psia). This design allows for a maximum vapor pressure of 215 psia at 100•F. The discharge piping for pumps and compressors is currently designed to 350 psi and vapor piping is designed for 250 psi according to NFPA 58, 1998 Edition, 3-2.10.2. The minimum temperatures are determined by the steel used in the design of the storage tank and the piping. Liquid propane (if released at atmospheric pressure) can refrigerate steel pipes and tanks down to temperatures of -44•F.

- • Another property of propane in its liquid form is its ability to greatly expand when heated. Therefore, NFPA 58, 1998 Edition sets the maximum filling capacity of large tanks in Table 4-4.2.2(b) to avoid overfilling.

You must maintain equipment specifications for all equipment that is part of a covered process, including your bulk storage tank(s), piping, pressure relief valves, hydrostatic relief valves, emergency shutoff valves, temperature, pressure and level gauges, valves, pumps, compressors, and hoses. Specifications for your bulk propane storage tank(s) are provided on the nameplate attached to the tank. If you do not have the information, obtain it from your vendor and keep all such information on file.

You must document the codes and standards you used to design and build your propane facility and that you follow to operate it. These codes will probably include the electrical and building codes that you must comply with under state or local laws. Your equipment vendors will be able to provide you with information on the codes they comply with for their products.

The equipment specifications and lists of standards and codes will probably ensure that your process is designed in compliance with recognized and generally good engineering practices.

See the EPA and OSHA guidance documents cited above for more information on the elements that must be documented as part of your safety information.

**Process Hazard Analysis.** You are required to conduct a process hazard analysis (PHA) to identify the hazards associated with your equipment and propane, the possible malfunctions of equipment or human errors that could cause a release, the safeguards needed to control hazards or prevent malfunctions or errors, and any steps needed to detect or monitor releases. If you are required to comply with NFPA-58, your review can focus on whether you are in compliance with that standard. You may need to consider external events as well as internal failures. If you are in an area subject to earthquakes, hurricanes, or floods, you should examine whether your system would survive these natural events without releasing propane. You should consider the potential impacts of lightning strikes and power failures. If your equipment could be hit by vehicles, you should examine the consequences of that. If you have anything near the process that could burn, ask yourself what would happen if the fire affected the propane tanks or equipment. See the EPA and OSHA guidance documents cited above for more information on analysis techniques and issues to be considered. Appendix C provides a checklist that you can use.

When you complete the review, you must document the results and ensure and document that any problems are addressed in a timely manner.

**Operating Procedures.** Written operating procedures describe the tasks you or your operators must perform, safe process operating parameters that must be maintained, and safety precautions for operations and maintenance activities. These procedures tell you or your employees how to work safely every day. Applicable portions of the National Propane Gas Association (NPGA) Certified Employee Training Program and compliance with certain NPGA Safety Bulletins can be used to meet this requirement. Other programs may be available that will also be acceptable. See the EPA and OSHA guidance documents cited above for more information on the elements that must be included in the procedures. A list of NPGA documents that may meet these requirements is included in Appendix C.

**Training.** You must ensure that any employee presently operating a process and any employee newly assigned to a covered process have been trained or tested competent in the operating procedures that pertain to their duties. For those employees already operating a process on June 21, 1999, you may certify in writing that the employee has the required knowledge, skills, and abilities to safely carry out the duties and responsibilities as provided in the operating procedures. You are not required to provide a specific amount or type of training. You should develop a training approach that works for you. The NPGA's Certified Employee Training Program and any training you do to meet DOT requirements may satisfy this requirement. You must provide refresher training. You must determine the frequency of refresher training in consultation with any affected employees, but you must provide refresher training at least once every three years. Training must be documented.

**Maintenance.** You must prepare and implement procedures to maintain the on-going mechanical integrity of your propane equipment. You may use procedures or instructions provided by equipment vendors or procedures in Federal or state regulations or industry codes as the basis for maintenance procedures. You must also train maintenance workers in these procedures (if a contractor maintains your equipment, the contractor's employees should be trained as well). NPGA's Certified Employee Training Program covers many of the maintenance procedures for your propane equipment.

You must establish a schedule for inspecting and testing equipment associated with your propane storage facility. You may obtain recommendations from manufacturers, vendors, or trade associations. You should, however, use your own experience as a basis for examining any schedules you obtain from others. Many things may affect whether a schedule is appropriate. The manufacturer may assume a constant rate of use. If your actual rate of use (e.g., the amount of propane pumped per hour) varies considerably, the variations may cause additional wear on the equipment. Extreme weather conditions may also increase wear on equipment.

If you have workers that use your propane facility, talk with them as you prepare or adopt these procedures and schedules. If their experience indicates that equipment fails more frequently than the manufacturer expects, you should adjust the inspection schedule to reflect that experience.

See the EPA and OSHA guidance documents cited above for more information on these requirements. A list of NPGA training that applies to maintenance procedures and a maintenance inspection checklist are included in Appendix C.

**Management of Change.** You must prepare and implement procedures to manage changes, except replacements in kind. The purpose of this requirement is to ensure that changes you make do not create any new hazards and that your employees and contractors are informed of the changes and trained in any new procedures that are needed prior to start-up. See the EPA and OSHA guidance documents cited above for more information on these requirements.

**Pre-Startup Review**. Whenever you install a new covered process or modify an existing one, you must perform a pre-startup review to ensure that the equipment is in accordance with the design specifications, that all procedures related to the equipment are up-to-date, and that employees have been trained in the operation of the new or modified process. See the EPA and OSHA guidance documents cited above for more information on these requirements.

**Compliance Audits.** At least every three years, you must certify that you have evaluated compliance with EPA's requirements for the prevention program for each covered process. At least one person who conducts the audit must be knowledgeable about the process. You must develop a report of the audit's findings, determine and document an appropriate response to each finding, and document that you have corrected all deficiencies. You must retain compliance audit reports for five years.

**Incident Investigation.** You must investigate each incident that resulted in, or could have resulted in, a "catastrophic" release of propane. A catastrophic release is one that presents an imminent and substantial endangerment to public health and the environment. You must start the investigation no later than 48 hours after the accident. You must create a report on the accident that includes, at least, the date of the accident and the date the investigation began, a description of the accident, the factors that contributed to the accident, and any recommendations that resulted from the investigation. You must address the recommendations and share the findings with any employees whose jobs are affected by the findings. Investigation reports must be retained for five years.

The NPGA "LP-Gas Safety Handbook," and NPGA bulletin #202-93 "After Accident Procedure" may help you comply with this requirement.

**Employee Participation.** You must develop a written plan to implement the requirements for employee participation. See the OSHA guidance document cited above for more information on this requirement.

**Hot Work Permit.** You must issue hot work permits for hot work conducted near your covered process. See the OSHA guidance document cited above for more information on this requirement.

**Contractors**. If you use contractors to perform maintenance, renovation, or specialty work on or near a covered process, you must implement a program to ensure that your contractors are qualified to perform the work and are informed of the potential hazards. See the OSHA guidance document cited above for more information on these requirements.

Table 4 summarizes the ways that propane facilities can comply with some of these requirements.

**TABLE 4**
**WAYS TO COMPLY WITH SOME PREVENTION ELEMENTS**

| Program 3 Prevention Element | How a Propane Facility Can Meet This Requirement |
|---|---|
| Safety Information | - Maintain Material Safety Data Sheets on propane<br>- Use the information provided in this guidance<br>- Document NFPA-58 information<br>- Maintain propane equipment vendor-supplied information<br>- Maintain records on electrical and building codes followed |
| Hazard Review | - Use checklist in Appendix C<br>- Review compliance with NFPA-58 |
| Operating Procedures | - Implement NPGA Certified Employee Training Program<br>- Comply with NFPA-58<br>- Comply with NPGA safety bulletins<br>- Use written operating procedures for propane systems |
| Training | - Implement NPGA Certified Employee Training Program.<br>- Document training done to meet DOT requirements<br>- Document training done to comply with NFPA-58<br>- Comply with NPGA Safety Bulletins<br>- Provide refresher training at least every three years |
| Maintenance | - Implement NPGA Certified Employee Training Program<br>- Use checklist in Appendix C<br>- Establish a maintenance and testing schedule<br>- Document inspections and maintenance done by equipment vendors |
| Compliance Audits | - Conduct and document a compliance audit every three years; respond to each finding, and document that you have corrected any deficiencies. |
| Incident Investigation | - Implement practices in NPGA's LP-Gas Safety Handbook<br>- Implement NPGA bulletin #202-93 "After Accident Procedure" |

## WHAT DO I HAVE TO DO FOR THE EMERGENCY RESPONSE PROGRAM?

If you have at least one Program 3 process at your facility, you may be required to implement an emergency response program, consisting of an emergency response plan, emergency response equipment procedures, employee training, and procedures to ensure the program is up-to-date. This requirement applies if your employees will respond to some releases involving propane. The emergency response section of EPA's rule allows you to decide first whether the employees will respond to an accidental release of propane and then what involvement the employees will have in the event of a release of propane. If you choose not to have employees respond, then you must coordinate response actions with the local fire department and have in place appropriate mechanisms to notify emergency responders when there is a need for a response.

Some propane storage facilities will rely on local responders to handle any accident. If you plan to have your employees respond to a propane release, you should consult EPA's *General Guidance for Risk Management Programs* to determine what you need to do to develop and implement an emergency response program.

## WHAT DO I HAVE TO DO FOR MY RMP?

The RMP for a Program 3 process will include the same sections covered in the Program 1 process, plus a report on the alternative release scenario and the report on the prevention program. Except for the executive summary, the RMP consists of names, numbers, and check-off boxes. If you have more than one process, you still file only one RMP. If you have multiple Program 3 processes, but they all contain propane, you report only one worst-case scenario and one alternative scenario to cover all of them. (If you have multiple Program 1 processes, you must report a worst-case scenario for each Program 1 process to establish that the process is eligible for Program 1.)

If you have one Program 3 process, your RMP will include:

- • The executive summary (covering the alternative release scenario as well as worst-case)
- • Registration data
- • Worst-case and alternative release data
- • Five-year accident history (only if you've had any accidents to report)
- • Prevention program data
- • Emergency response data
- • The certification

A sample RMP and certification statement for a Program 3 propane facility is attached.

# SAMPLE RMP for PROGRAM 1 PROPANE STORAGE FACILITY
*(This sample RMP is for a fictitious facility named, "Smith Gas Company Terminal."*
*Any resemblance to any actual facility is accidental).*

## CERTIFICATION STATEMENT

Based on the criteria in 40 CFR 68.10, the distance to the specified endpoint for the worst-case accidental release scenario for the following processes is less than the distance to the nearest public receptor:

•• Wholesale LPG facility

Within the past five years, the process has had no accidental release that caused offsite impacts provided in the risk management program rule (40 CFR 68.10(b)(1)). No additional measures are necessary to prevent offsite impacts from accidental releases. In the event of fire, explosion, or a release of a regulated substance from the process, entry within the distance to the specified endpoint may pose a danger to public emergency responders. Therefore, public emergency responders should not enter this area except as arranged with the emergency contact indicated in the RMP. The undersigned certifies that, to the best of my knowledge, information, and belief, formed after reasonable inquiry, the information submitted is true, accurate, and complete.

| | |
|---|---|
| *William R. Smith* | William R. Smith |
| Signature | Print Name |
| | |
| Company Owner | 6/21/99 |
| Title | Date |

## EXECUTIVE SUMMARY

***The accidental release prevention and emergency response policies at your facility***: This facility complies with NFPA-58 requirements for LP-Gas storage, and it is our policy to adhere to all applicable federal, state, and local laws. If an emergency were to occur, it is our policy to notify the Garvin County Fire Department and request that they respond to the emergency.

***A description of your facility and the regulated substances handled***. This facility wholesales LPG, supplying retail outlets, primarily gas stations and cylinder filling operations at cooperatives. The facility consists of a single 18,000-gallon tank, which we fill to between 85% and 88% capacity.

***The worst-case release scenario***. Our worst-case scenario is failure of the 18,000-gallon storage tank when filled to the greatest amount allowed (88% at 60F), resulting in a vapor cloud explosion. Since this facility is located in a relatively remote, unoccupied area, the worst-case scenario would not affect anyone beyond our property.

***The general accidental release prevention program and chemical-specific prevention steps***. This facility complies with EPA's accident prevention rule and all applicable state and local codes and regulations. The propane system is designed, installed, and maintained in accordance with NFPA-58 and state law.

***Five-year accident history***. We have never had an accident involving propane that caused deaths, injuries, property or environmental damage, evacuations, or shelterings in place.

***The emergency response program***. In the event of an emergency involving our propane system, it is our policy to notify the Garvin County Fire Department and request that they respond to the emergency. We have discussed this policy with the fire department; members of the fire department have inspected our propane system.

January 27, 2000

*Planned changes to improve safety.* None.

## 1. REGISTRATION

1.1 Source Identification

    1.1.a. Facility Name:     **Smith Gas Company**

    1.1.b. Parent Company #1 Name: N/A

    1.1.c. Parent Company #2 Name:

1.2. RMP Facility Identifier: [EPA will assign]

1.3. EPA Identifier:

1.4. Dun and Bradstreet Numbers (DUNS) N/A

    1.4.a. Facility DUNS:

    1.4.b. Parent Company #1 DUNS:

    1.4.c. Parent Company #2 DUNS:

1.5 Facility Location Address

    a. Street     **42 Rural Rt 7**

    b. Street - Line 2:

    c. City: **Plainville**     d. State: **OK**     e. Zip Code:     **12345**     f. County:     **Garvin**

    g. Facility Latitude (degrees, minutes, and seconds):     **34**     **40**     **20**

    h. Facility Longitude (degrees, minutes, and seconds): **-097**     **21**     **06**

    i. Method for determining Lat/Long :     **I1 (interpolation, map)**

    j. Description of location identified by Lat/Long :     **AB**     **Storage Tank**

1.6 Owner/Operator

    a. Name:     **William R. Smith**

    b. Phone:     **(555) 555-5555**

    Mailing Address:

    c. Street 1:     **42 Rural Rt 7**

    e. City:     **Plainville**     f. State: **OK**     g. Zip: **12345**

1.7. Name and title of person responsible for RMP (part 68) implementation

    a. Name:     **William R. Smith**

b. Title:    **Company owner**

1.8. Emergency Contact

a. Name:    **William R. Smith**

b. Title:    **Company owner**

c. Phone:    **(555) 555-5555**

d. 24-hour phone:    **(555) 555-1111**    e. Ext. or PIN:

1.9. Other Points of Contact (Optional)

a. Facility or parent company e-mail address:

b. Facility public contact phone:    **(555) 555-5555**

c. Facility or parent company www homepage address:

1.10. LEPC (Optional):    **Garvin County LEPC**

1.11. Number of full-time employees (FTEs) On Site:    **1**

1.12. Covered by (select all that apply)

a. OSHA PSM:    **Yes**

b. EPCRA section 302:

c. CAA Title V Air Operating Permit ID:

1.13. OSHA Star or Merit Ranking: **No**

1.14. Last Safety Inspection Date: **12/07/97**

1.15. Last Safety Inspection Performed by (select one)    **Fire department**

1.16. Will this RMP involve Predictive Filing? **No**

1.17. Process Specific Information. For each covered process fill in the following chart. Use a separate sheet for each process

| | |
|---|---|
| Process Number: (optional to help you track) | 1 |
| Process Description: (optional to help you track) | **Storage tank** |
| a. Program Level: | 1 |
| b. NAICS Code(s): | 42272 |

January 27, 2000

| c. Chemical | c.1. Name: | c.2. CAS Number: | c.3. Quantity (lbs.): |
|---|---|---|---|
| | **Propane** | **74-98-6** | **67,000** |

## 4. FLAMMABLES: WORST CASE

4.1. Chemical Name **Propane**

4.2. Results based on (select one)

c. EPA's *RMP Guidance for Propane Storage Facilities* **Reference Tables or Equations**

4.3. Scenario: **Vapor Cloud Explosion**

4.4. Quantity released (lbs.) **67,000 pounds**

4.5. Endpoint Used: **1 psi**

4.6. Distance to endpoint (miles) **0.30 miles**

4.7. Residential population within distance to endpoint **0**

4.8. Public receptors within distance to endpoint (select all that apply)

a. Schools                    d. Prisons /Correctional facilities

b. Residences              e. Recreation areas

c. Hospitals                 f. Commercial/industrial areas

4.9. Environmental receptors within distance to endpoint (select all that apply)

a. National or state parks, forests, or monuments

b. Officially designated wildlife sanctuaries, preserves, or refuges

c. Federal wilderness area

4.10. Passive mitigation considered (select all that apply)

a. Dikes

b. Fire walls

c. Blast walls

d. Enclosures

e. Other (specify)

4.11. Graphics file name (Optional)

## 9. EMERGENCY RESPONSE

9.1. Emergency response (ER) plan

    a. Is facility included in the written community emergency response plan?    **No**

    b. Does facility have its own written emergency response plan?    **No**

9.2. Does facility ER plan include specific actions to be taken in response to accidental releases of regulated substance(s)?

9.3. Does facility ER plan include procedures for informing public and local agencies responding to accidental release?

9.4. Does facility ER plan include information on emergency health care?

9.5. Date of most recent review/update of facility ER plan

9.6. Date of most recent emergency response training for facility's employees

9.7. Local agency with which the facility ER plan or response activities are coordinated

    a. Name of agency    **Garvin County Fire Department**

    b. Phone number **(555) 555-1000**

9.8. Subject to (select all that apply)

    9.8.a. OSHA 1910.38

    9.8.b. OSHA 1910.120

    9.8.c. Clean Water Act/SPCC

    9.8.d. RCRA

    9.8.e. OPA-90

    9.8.f. State EPCRA rules/law

    9.8.g. Other (specify)

22

# SAMPLE RMP for PROGRAM 3 PROPANE STORAGE FACILITY
*(This sample RMP is for a fictitious facility named "Maryland LPG."
Any resemblance to any actual facility is accidental).*

## CERTIFICATION STATEMENT

To the best of the undersigned's knowledge, information, and belief formed after reasonable inquiry, the information submitted is true, accurate, and complete.

| | |
|---|---|
| ___*Mary L. Jones*___ | ___Mary L. Jones___ |
| Signature | Print Name |
| ___Company Owner___ | ___6/21/99___ |
| Title | Date |

## EXECUTIVE SUMMARY

***The accidental release prevention and emergency response policies at your facility***: This facility complies with NFPA-58 requirements for LP-Gas storage, and it is our policy to adhere to all applicable federal, state, and local laws. If an emergency were to occur, it is our policy to notify the Howard County Fire Department and request that they respond to the emergency. We also maintain a fire brigade on site to handle small emergencies.

***A description of your facility and the regulated substances handled***. This facility is a bulk terminal storing LPG for resale to retail propane outlets. The facility consists of four 30,000-gallon tanks and two 120,000-gallon tanks.

***The worst-case and alternative release scenarios.*** Our worst-case scenario is failure of one 120,000-gallon storage tank when filled to the greatest amount allowed (88% at 60F), resulting in a vapor cloud explosion. The resulting distance to the endpoint extends offsite, and public receptors are within the distance to the endpoint. Our alternative release scenario is a break in a 0.5-inch diameter pipe, leading to a 10-minute release and explosion. The resulting distance to the endpoint extends offsite, and public receptors are within the distance to the endpoint.

***The general accidental release prevention program and chemical-specific prevention steps.*** This facility complies with EPA's accident prevention rule, the OSHA PSM standard, and all applicable state and local codes and regulations. The propane system is designed, installed, and maintained in accordance with NFPA-58 and state law.

***Five-year accident history***. We have never had an accident involving propane that caused deaths, injuries, property or environmental damage, evacuations, or shelterings in place.

***The emergency response program.*** In the event of an emergency involving our propane system, it is our policy to notify the Howard County Fire Department and request that they help respond to the emergency. We also have a trained fire brigade. We have coordinated our response plans with the fire department; members of the fire department have inspected our propane system and conducted a joint exercise with us.

***Planned changes to improve safety.*** None.

## 1. REGISTRATION

1.1 Source Identification

    a. Facility Name: **Maryland LPG**

    b. Parent Company #1 Name: N/A

    c. Parent Company #2 Name:

1.2. RMP Facility Identifier: [EPA will assign]

1.3. EPA Identifier:

1.4. Dun and Bradstreet Numbers (DUNS) N/A

    a. Facility DUNS:

    b. Parent Company #1 DUNS:

    c. Parent Company #2 DUNS:

1.5 Facility Location Address

a. Street      **238 Main Street**

b. Street - Line 2:

c. City: **Odenton**      d. State: **MD**      e. Zip Code:    **21873**    f. County:      **Howard**

g. Facility Latitude (degrees, minutes, and seconds):    **39**    **11**    **15**

h. Facility Longitude (degrees, minutes, and seconds): **-076**    **50**    **10**

i. Method for determining Lat/Long :      **I1 (interpolation, map)**

j. Description of location identified by Lat/Long :      **AB**      **Administrative Building**

1.6 Owner/Operator

a. Name:      **Mary L. Jones**

b. Phone:      **(410) 777-1234**

Mailing Address:

c. Street 1:      **238 Main St.**

e. City:      **Odenton**      f. State: **MD**    g. Zip: **21873**

1.7. Name and title of person responsible for RMP (part 68) implementation

    a. **Mary L. Jones**

    b. **Company Owner**

January 27, 2000

1.8. Emergency Contact

a. Name: **Mary L. Jones**

b. Title: **Company owner**

c. Phone: **(410) 777-1234**

d. 24-hour phone: **(410) 777-4321**  e. Ext. or PIN:

1.9. Other Points of Contact (Optional)

a. Facility or parent company e-mail address:

b. Facility public contact phone: **(410) 777-1234**

c. Facility or parent company www homepage address:

1.10. LEPC (Optional): **Howard County LEPC**

1.11. Number of full-time employees (FTEs) On Site: **6**

1.12. Covered by (select all that apply)

    a. OSHA PSM: **Yes**

    b. EPCRA section 302:

    c. CAA Title V Air Operating Permit ID:

1.13. OSHA Star or Merit Ranking: **No**

1.14. Last Safety Inspection Date: **10/19/98**

1.15. Last Safety Inspection Performed by (select one) **Fire department**

1.16. Will this RMP involve Predictive Filing? **No**

1.17. Process Specific Information. For each covered process fill in the following chart. Use a separate sheet for each process.

| | |
|---|---|
| Process Number: (optional to help you track) | 1 |
| Process Description: (optional to help you track) | **Storage Tanks** |
| a. Program Level: | 3 |
| b. NAICS Code(s): | **42271 Petroleum Bulk Terminal** |

| c. Chemical | c.1. Name: | c.2. CAS Number: | c.3. Quantity (lbs.): |
|---|---|---|---|
| | **Propane** | **74-98-6** | **1,331,000** |

## 4. FLAMMABLES: WORST CASE

4.1. Chemical Name      **Propane**

4.2. Results based on (select one)

     c. EPA's *RMP Guidance for Propane Storage Facilities* **Reference Tables or Equations**

4.3. Scenario: **Vapor Cloud Explosion**

4.4. Quantity released (lbs.)      **120,000 pounds**

4.5. Endpoint Used: **1 psi**

4.6. Distance to endpoint (miles)      **0.60 miles**

4.7. Residential population within distance to endpoint      **200**

4.8. Public receptors within distance to endpoint (select all that apply)

a. Schools • •      d. Prisons /Correctional facilities

b. Residences    • • •      e. Recreation areas • •

c. Hospitals • •      f. Commercial/industrial areas • •

4.9. Environmental receptors within distance to endpoint (select all that apply)

     a. National or state parks, forests, or monuments

     b. Officially designated wildlife sanctuaries, preserves, or refuges

     c. Federal wilderness area

4.10. Passive mitigation considered (select all that apply)

     a. Dikes

     b. Fire walls

     c. Blast walls

     d. Enclosures

     e. Other (specify)

4.11. Graphics file name (Optional)

**5. FLAMMABLES: ALTERNATIVE RELEASES** [Program 2 processes only]

5.1. Chemical Name     **Propane**

5.2. Results based on: **EPA's** *RMP Guidance for Propane Storage Facilities* **Reference Tables or Equations**

5.3. Scenario     **Vapor cloud explosion**

5.4. Quantity released (lbs.)     **4,738**

5.5. Endpoint used (select one)   **1 psi**

5.6. Distance to endpoint (miles)   **0.10**

5.7. Residential population within distance to endpoint     **14**

5.8. Public receptors within distance to endpoint (select all that apply)

　　a. Schools　　　　　　　　　d. Prisons /Correctional facilities

　　b. Residences　• • •　　　　e. Recreation areas

　　c. Hospitals　　　　　　　　f. Commercial/industrial areas • •

5.9. Environmental receptors within distance to endpoint

　　a. National or state parks, forests, or monuments

　　b. Officially designated wildlife sanctuaries, preserves, or refuges

　　c. Federal wilderness area

5.10. Passive mitigation considered (select all that apply)

　　a. Dikes

　　b. Fire walls

　　c. Blast walls

　　d. Enclosures

　　e. Other (specify)

5.11. Active mitigation considered (select all that apply)

　　a. Sprinkler system

　　b. Deluge system

　　c. Water curtain

　　d. Excess flow valve

　　e. Other (specify)

January 27, 2000

5.12. Graphics file name (Optional)

## 7. PREVENTION PROGRAM - PROGRAM 3

For each process or process unit:

7.1. NAICS Code for process: 42271

| 7.2. Chemical name(s): | **Propane** |
|---|---|

7.3. Safety information

    Date of most recent review/revision of safety information    **02/04/97**

7.4. Hazard review

    a.1 Date of completion of most recent PHA/update    **02/04/98**

    a.2  Technique used    **Checklist**

    b. Expected date of completion of any changes resulting from the PHA    **02/04/98**

    c. Major hazards identified (select at least one)

| Toxic release | Overpressurization | • • • | Earthquake |
|---|---|---|---|
| Fire • • • | Corrosion | | Floods |
| Explosion • • • | Overfilling | • • • | Tornado |
| Runaway reaction | Contamination | | Hurricanes |
| Polymerization | Equipment Failure | • • • | Other |

           Loss of cooling, heating, electricity, instrument air

    7.4.d. Process controls in use (select at least one)

| Vents • • • | Emergency air supply | |
|---|---|---|
| Relief valves • • • | Emergency power | |
| Check valves | Backup pump | |
| Scrubbers | Grounding equipment | • • • |
| Flares | Inhibitor addition | |
| Manual shutoffs • • • | Rupture disks | |
| Automatic shutoffs • • • | Excess flow device | • • • |

January 27, 2000

Interlocks                    Quench system

Alarms and procedures    • • •    Purge system

Keyed bypass                  Other:    **Breakaway couplings**

7.4.e. Mitigation systems (select all that apply)

Sprinkler system        • • •      Deluge system

Dikes                         Water curtain

Fire walls                    Enclosure

Blast walls                   Neutralization

                              Other (specify)

7.4.f. Monitoring/detection systems (select all that apply)

Process area detectors        Other (specify)

Perimeter monitors

7.4.g. Changes since last PHA update (select all that apply)

Reduction in chemical inventory          Installation of perimeter monitoring systems

Increase in chemical inventory           Installation of mitigation systems

Change in process parameters             None required/recommended

Installation of process controls    • • •    Other (specify)

Installation of process detection systems

7.5. Date of most recent review/revision of operating procedures    **03/01/97**

7.6. Training

a. Date of most recent review/revision of training programs    **03/01/97**

b. Type of training provided (select at least one)

Classroom                     On the job      • • •

Other (specify)

c. Type of competency test used (select at least one)

Written test                  Observation     • • •

Oral test                     Other (specify)

Demonstration    • • •

7.7. Maintenance

    a. Date of most recent review/revision of maintenance procedures    **04/01/97**

    b. Date of most recent equipment inspection/test    **10/19/98**

    c. What equipment inspected/tested    **Propane tanks, valves, and piping**

7.8 Management of Change

    a. Date of the most recent change that triggered management of change procedure:

    b. Date of the most recent review or revision of management of change procedures: **6/14/98**

7.9  Pre-Startup Review

    The date of the most recent pre-startup review:

7.10. Compliance audits

    a. Date of most recent compliance audit

    b. Expected date of completion of any changes resulting from the compliance audit

7.11. Incident investigation:

    a. Date of most recent incident investigation

    b. Expected date of completion of any changes resulting from the investigation

7.12  Employee participation plan

    Date of the most recent review or revision of the employee participation plan: **4/16/99**

7.12 Hot work permit procedures

    Date of the most recent review or revision of the hot work permit procedures: **24/12/97**

7.13 Contractor safety procedures

    a. Date of the most recent review or revision of the contractor safety procedures: **4/16/99**

    b. The date of the most recent review or revision of contractor safety performance: **4/16/99**

## 9. EMERGENCY RESPONSE

9.1. Emergency response (ER) plan

    a. Is facility included in the written community emergency response plan?    **Yes**

    b. Does facility have its own written emergency response plan?    **Yes**

9.2. Does facility ER plan include specific actions to be taken in response to accidental releases of regulated substance(s)? **Yes**

9.3. Does facility ER plan include procedures for informing public and local agencies responding to accidental release? **Yes**

9.4. Does facility ER plan include information on emergency health care? **Yes**

9.5. Date of most recent review/update of facility ER plan    **11/06/98**

9.6. Date of most recent emergency response training for facility's employees **3/14/99**

9.7. Local agency with which the facility ER plan or response activities are coordinated

    a. Name of agency    **Howard County Fire Department**

    b. Phone number **(410) 123-4567**

9.8. Subject to (select all that apply)

    9.8.a. OSHA 1910.38

    9.8.b. OSHA 1910.120

    9.8.c. Clean Water Act/SPCC

    9.8.d. RCRA

    9.8.e. OPA-90

    9.8.f. State EPCRA rules/law

    9.8.g. Other (specify)

## APPENDIX A
## HOW DO I DETERMINE IF MY SEPARATE (NON-INTERCONNECTED) PROPANE TANKS ARE CO-LOCATED?

January 27, 2000

# APPENDIX A
# HOW DO I DETERMINE IF MY SEPARATE (NON-INTERCONNECTED) PROPANE TANKS ARE CO-LOCATED?

For separate, above-ground, non-interconnected propane tanks, one way to determine whether the tanks constitute a single process (i.e., in rule language are "co-located") is to determine whether they are close enough together for a vapor cloud explosion resulting from the release of the total contents of one tank to cause the catastrophic failure of an adjacent tank. Table A-1 indicates estimated separation distances based on this method.

## TABLE A-1
## ESTIMATED SEPARATION DISTANCES FOR CONSIDERING NON-INTERCONNECTED PROPANE TANKS AS SEPARATE PROCESSES

| Tank Capacity (gal) | Tank Confinement | Estimated Separation Distance (ft) |
|---|---|---|
| 500 | Partial - High | 61 |
| 500 | Low | 41 |
| 1000 | Partial - High | 76 |
| 1000 | Low | 51 |
| 2000 | Partial-High | 96 |
| 2000 | Low | 64 |

The estimated separation distances in Table A-1 corresponding to "Partial - High" confinement are based on a TNT-equivalent yield factor of 10 percent. This is the same yield factor specified in the RMP rule for conducting flammable gas worst-case scenario modeling using TNT-equivalent methods (readers should note that the distances in Table A-1 are not intended as worst-case endpoint distances). This yield factor is appropriate for vapor cloud explosions occurring in areas with numerous obstructions, such as in pipe racks, between stacks of crates or pallets, or near other closely spaced structures[1]. Table A-1 distances corresponding to "Low" confinement may be used if your propane tanks are located outdoors[2] in relatively flat, open terrain with few structures nearby (these separation distances are based on a TNT-equivalent yield factor of 3%). Otherwise, the distances corresponding to "Partial-High" confinement should be used to give reasonably conservative separation distance estimates.

---

[1] Vapor cloud explosions occurring in congested or confined spaces are generally stronger due to higher flame speeds resulting from turbulence produced in unburned gases expanding ahead of the propagating flame front.

[2] NFPA-58 generally requires propane tanks to be located outside of buildings.

To determine whether adjacent, non-interconnected propane tanks constitute a single process, measure the linear distance between the tanks and compare it to the estimated separation distance in Table A-1 for tanks of that size and with that degree of confinement. Tanks that are separated by less than the estimated separation distance in Table A-1 should generally be considered a single process and their quantities added together to determine if the process exceeds the RMP threshold of 10,000 pounds. Individual tanks of propane which are separated by at least the estimated separation distance in Table A-1 for tanks of that size and with that degree of confinement may generally be considered as separate processes and thus not subject to the RMP regulation.

The distances provided in the Table above should be treated as estimates. Other methods of determining separation distances may provide more accurate estimates and you are free to use other reasonable methods if you choose. Also, the separation distances provided here should be considered in relation to any unique circumstances at your site, such as topography, the presence of structures and obstacles, the presence of blast-mitigation features, and other site-specific factors. For example, if engineered blast-barriers are located between tanks, or if your tanks are underground, smaller separation distances may be appropriate[3]. On the other hand, topographical features, such as a depression between two tanks where heavier-than-air propane vapor might collect, may warrant the use of larger separation distances. Additionally, you should evaluate whether there are potential events other than vapor cloud explosions that could reasonably be expected to cause multiple tanks to fail, even at distances larger than those in Table A-1. EPA believes that for most properly designed outdoor propane installations, such events are extremely unlikely, and that separating tanks by at least the appropriate distance in Table A-1 is sufficient to establish separate processes. However, you should evaluate your own circumstances and maintain documentation to support your determination.

Some readers of this guidance may conclude that it will be easier to re-locate one or more of their propane tanks to conform to the distances provided above rather than taking the steps necessary to comply with the regulation. This is certainly permitted, and in some cases may be an appropriate risk-reduction measure. However, when taking such actions, you should be careful to maintain sufficient separation distances between your propane tanks and nearby buildings, roads, and public receptors. EPA recommends consulting NFPA-58 or other applicable codes or standards to identify such distances.

---

[3] NFPA-58 generally recommends against installing fire walls, fences, earth or concrete barriers, and other similar structures around or over non-refrigerated LP-gas containers. There are some exceptions. For example, the standard permits such structures to partially enclose LP-gas containers, if the structure is designed in accordance with a sound fire protection analysis.

**APPENDIX B**
**RELEASE CALCULATIONS**

February 24, 1999

# APPENDIX B - RELEASE CALCULATIONS

The endpoints that can be considered for the alternative release scenario are explosion damage, radiant heat effect and lower flammability limit vapor concentration. Review these potential alternative scenarios and pick one or more that best describes a scenario that is more likely to occur than the worst-case scenario and that has an endpoint which is beyond your facility boundary. The following are possible alternative release scenarios that use a vapor cloud explosion and an overpressure of 1 psi.

## PULL-AWAY: (ACTIVE MITIGATION DEVICES PERFORM AS DESIGNED)

An alternative scenario for a propane storage facility may be a hose rupture caused by a pull-away. A pull-away can occur if the driver fails to remove the hoses between the storage tank and the transfer vehicle (bobtail or transport) before moving the vehicle. In this case, the analysis considers the failure of a 25 foot length of unloading hose, 4" in diameter. The active mitigation devices are assumed to work as designed, limiting the release to the contents of the hose.

Volume of hose = cross sectional area of hose x length of hose
Volume of hose = • •× (4 in / 2)$^2$ × 1 ft$^2$ / 144 in$^2$ × 25 ft = 2.182 ft$^3$
$W_f$ = 2.182 ft$^3$ × .504 × 62.37 pounds/ft3 = 68.59 pounds of propane

There are three methods described in EPA's OCA that you can use to determine the distance to the endpoint for an alternative scenario. Two of those methods are used here as an example.

## PULL-AWAY EXPLOSION

You can use equation C-1 from EPA's OCA to calculate the distance exposed to greater than 1 psi overpressure. This method results in a distance to endpoint of 175 feet and the endpoint is 1 psi overpressure.
$$D = 17 ( 0.1 \times W_f \times HC_f / HC_{TNT} )^a$$

Where,
$W_f$    is the weight of propane in kilograms in the vapor cloud,
$Hc_f$    is the heat of combustion of propane from Appendix C of EPA's OCA,
$Hc_{TNT}$  is the heat of combustion of TNT from Appendix C of EPA's OCA, and
D    is the distance in meters from the explosion where overpressure exceeds 1 psi.

$D = 17 (.1 \times Wf \times 46333/4680)^a = 17 (.1 \times 68.6/2.2046 \times 46333/4680)^a = 53.29$ meters =
$D = 174.8$ feet

If you use this scenario as your alternative release scenario, the distance to endpoint is 175 feet and the endpoint used is 1 psi overpressure.

## PROCESS PIPING BREAKS

Liquid propane pipe failures: ½", 1", 2", 3", 4" and 6". (Causes: liquid expansion [failure of hydrostatic valves], a pull-away [failure of breakaway devices[4]], collision, corrosion.) For long pipes, where the length to diameter ratio is significantly greater than one, flashing (i.e., conversion of gaseous propane to liquid) in the discharge pipe results in a two-phase release. You can estimate the amount released from a line shear using the following equation:

$$QR = 0.0853 \times HA \times F \times L / [V_{gl} \times (T \times C_p)^{0.5}]$$

Where: QR   is the release rate in pounds per minute.
      HA   is the cross-sectional area of the pipe in square inches.
      F   is the frictional factor, which is 0.85 for pipe with a length to diameter ratio of 50.
      L   is the heat of vaporization, which is 426,000 joules per kilogram for propane.
      $V_{gl}$   is the specific volume of gas minus liquid in $m^3$/kg, which is 0.056 for propane.
      T   is the temperature in Kelvins (assumed to be 298).
      $C_p$   is the heat capacity, which is 2,400 joules/kilogram/Kelvin for propane.
      0.0853  is a units conversion factor.

QR can be calculated using the values in the table below for HA. Now Q (the release quantity) can be calculated for a 10-minute release, and D (the distance to the endpoint in feet) can be calculated using equation C-1 from EPA's OCA. Note that Q cannot be larger than the amount of propane contained in your storage tank (this should be considered especially for 6" pipes).

For a 1" pipe:

QR = 512.5 pounds per minute

Q = QR × 10 minutes = 5,125 total pounds released (Q = $W_f$)

$$D = 17 ( 0.1 \times W_f \times HC_f / HC_{TNT})^a = 17 \times (0.1 \times 5,125 / 2.2046 \times 46333 / 4680)^a = 224$$

D = 224 meters = 735 feet = 0.1 mile

If you use this scenario as your alternative release scenario, the distance to endpoint is 735 feet (or about 0.1 mile) and the endpoint used is 1 psi overpressure.

---

[4] See NFPA 58, 1998 Edition, Section 3-9.4.2. A breakaway device is used when dispensing LPG. It is designed so that the breakaway device fails (instead of the piping or hose) allowing other safety devices to retain the fuel and limit the spill. Failure of the breakaway device means that the breakage occurs somewhere else rather than at the breakaway device during a pull away, thus negating the effectiveness of some or all of the other safety devices.

**TABLE B-1**
**POTENTIAL DISTANCES TO ENDPOINT FOR PIPE RELEASES**
**(for 1 psi Overpressure)**

| Pipe Size (inches) | HA (inches$^2$) | QR (pounds/minute) | Q Total Amount Released after 10 minutes (pounds) | D (Miles) |
|---|---|---|---|---|
| 0.50 | 0.20 | 128 | 1,281 | 0.1 |
| 1 | 0.79 | 512 | 5,125 | 0.1 |
| 2 | 3.14 | 2,050 | 20,499 | 0.2 |
| 3 | 7.07 | 4,612 | 46,122 | 0.3 |
| 4 | 12.57 | 8,199 | 81,995 | 0.4 |
| 6 | 28.27 | 18,449 | 184,488 | 0.5 |

## OVERFILL STORAGE TANK - RELIEF VALVE LIFTS

A safety relief valve lifts (causes: overpressure or overfilling [failure of your level indicator and administrative procedures]).

A four-valve multiport relief valve has a flow capacity of 27,750 SCFM/Air at 300 psi with three of the four valves lifting. Estimate the amount of propane discharged from one port over 5 minutes and the distance to endpoint if the vapor cloud finds a source of ignition and explodes. The density of air is 12.39 ft$^3$/lb and the density of propane is 8.66 ft$^3$/lb. To convert SCFM $_{(air)}$ to SCFM $_{(propane)}$ multiply by 0.808.

The release rate from one port is 27,950 SCFM $_{(air)}$/ 3 = 9,250 SCFM (air)

$Q_{(propane)} = Q_{(air)} \times 0.808 / 8.66$ ft$^3$/lb = 863 lbs/min $\times$ 5 min = 4,315lbs = 1957.4 kg

D = 17 ( 0.10 $\times$ 1957.4 $\times$ 46,333/4680)$^a$ = 211.9 meters $\times$ 3.281 ft/meter = 695.2ft / 5280ft/mile = 0.13 miles.

**TABLE B-2**
**RELIEF VALVE DISCHARGES (FOR 1 PSI OVERPRESSURE)**

| Relief Valve Size (inches) | Capacity SCFM of Air | Release Rate (lbs/min) Propane | Duration (min) | Amount Released (lbs) | Distance to Endpoint (Miles) |
|---|---|---|---|---|---|
| Multiport 2½" | 9,250 | 863 | 5 | 4,315 | 0.1 |
| Dual Port 1¼" | 5,250 | 490 | 5 | 2,449 | 0.08 |
| 2½" | 10,390 | 969 | 5 | 4,847 | 0.1 |
| 1½" | 6,080 | 567 | 5 | 2,836 | 0.1 |
| 1¼" | 5,280 | 492 | 5 | 2,463 | 0.1 |
| 1 " | 3,340 | 312 | 5 | 1,558 | 0.09 |

# APPENDIX C
# CHECKLISTS AND OTHER DOCUMENTS

March 4, 1999

# PROCESS SAFETY INFORMATION

| Process Safety Information Worksheet: Propane Storage ||
| :--- | :--- |
| **Item** | **Description** |
| MSDS Propane | |
| Maximum Intended Inventory | |
| Nominal Water Capacity of Largest Tank | |
| Temperature | Upper:<br>Lower: |
| Pressure | Upper:<br>Lower: |
| Flow Rate | Loading:<br>Unloading: |
| Vapor Piping | |
| Liquid Piping | |
| Safety Relief Valves<br><br>RV1<br>RV2<br>RV3<br>RV4 | |
| Internal Valve | |
| Excess Flow Valve<br>EFV1<br>EFV2<br>EFV3<br>EFV4 | |
| Emergency Shutoff Valve<br>ESV1<br>ESV2 | |

| Item | Description |
|---|---|
| Hydrostatic Relief Valve<br>HSV1<br>HSV2<br>HSV3<br>HSV4<br>HSV5<br>HSV6 | |
| Pump 1 | |
| Pump 2 | |
| Compressor 1 | |
| Compressor 2 | |
| Vaporizer | Type:          Sz.                    Flow: |
| Strainer | |
| Check valves | |
| Sightflow Indicator | |
| Tank Level Indicator | |
| Tank Temperature Indicator | |
| Tank Pressure Indicators | |
| Other Temperature Indicators | |
| Other Pressure Indicators | |
| **Design Codes** | |
| State or Local Codes | |
| Piping Design | |
| Tank Design | ASME NB#                    State ID# |
| Vaporizer Design | ASME NB#                    State ID# |
| **Date of Most Recent Revision** | **Revised By:** |
| | |

## PROPANE STORAGE FACILITY CHECKLIST[5,6]

Answer the questions below by indicating "Yes", "No" or "N/A" (for not applicable). "No" responses require further comment and a projected completion date for correcting the deficiency.

| Siting | Yes/No/NA | Comments |
|---|---|---|
| 1. Does the arrangement of your fixed storage tanks conform with the minimum distances allowed in Table 3-2.2.2 of NFPA 58, 1998 Edition? | | |
| 2. Are your fixed storage tanks separated from any oxygen or hydrogen storage by the minimum distances given in Table 3-2.2.7(f) of NFPA 58, 1998 Edition? | | |
| 3. Are your transfer points separated from the exposure points by the minimum distances given in Table 3-2.3.3 of NFPA 58, 1998 Edition? | | |
| **Piping, Equipment & Container Appurtenances** | **Yes/No/NA** | **Comments** |
| 1. Is your storage facility designed according to ASME code for pressure vessels? <br><br> Fixed Storage Tanks ASME? <br><br> Vaporizers ASME? | | |
| 2. Is the pressure rating of your storage tanks appropriate for the product in service? <br><br> Storage Tanks? <br><br> Vaporizers? | | |
| 3. Is the stored product properly identified? | | |
| 4. On installations with multiple tanks, are the elevations of your storage tanks arranged to prevent unintentional overfilling of the lowest container? | | |

---

[5] Completing the Propane Storage Facility Hazard Review Checklist does not guarantee that your facility is in complete compliance with NFPA 58, 1998 Edition.

[6] The Propane Storage Facility Hazard Review Checklist is based on NFPA 58, 1998 Edition. Over the years, various changes in NFPA 58 have been made to reduce the probability of a propane release. You should carefully review any change to NFPA 58 since your facility was constructed and consider appropriate changes to ensure the safety of your facility. Document the actions that you take. You should be ready to explain any differences between the version of NFPA 58 that you used to construct your facility and the current version.

| Piping, Equipment & Container Appurtenances | Yes/No/NA | Comments |
|---|---|---|
| 5. On installations with stairways or ladders, are they well anchored, supported and of slip proof construction? | | |
| 6. On installations with stairways or ladders, are railings provided and in good condition? | | |
| 7. On installations with stairways or ladders, are catwalks provided so personnel need not walk on any portion of the container? | | |
| 8. Is your piping designed according to NFPA 58, 1998 Edition, Section 3-2.10?<br><br>Are your pump and compressor discharge and liquid transfer lines suitable for a working pressure of 350 psi?<br><br>Is your vapor piping suitable for a working pressure of 250 psi?<br><br>On installations with vaporizers, are your vaporizers designed according to 2-5.4.2 or 2-5.4.3 or 2-5.4.4 and 2-5.4.5 or 2-5.4.6 or 2-5.4.7 of NFPA 58, 1998 Edition? | | |
| 9. Is the relief capacity of your pressure relief devices:<br><br>For fixed storage tanks, designed according to Sections 2-3.2 and 3-2.5 or 3-2.6 of NFPA 58, 1998 Edition?<br><br>On installation with vaporizers, are your vaporizers designed according to 2-5.4.5 or 2-5.4.6 or 2-5.4.7 of NFPA 58, 1998 Edition? | | |
| 10. Is the capacity of your pressure relief devices designed according to 2-3.2 and 3-2.5 or 3-2.6 of NFPA 58, 1998 Edition?<br><br>Have your relief devices been tested or replaced every ten years according to the good practice recommended by Section E-2.3.2 of NFPA 58, 1998 Edition? | | |
| 11. Do you have appropriate level gauges, temperature indicators, and pressure gauges installed on your fixed ASME storage tanks as specified in 2-3.3.2(b), 2-3.3.3, 2.3.4, 2.3.5 of NFPA 58, 1998 Edition? | | |

| Piping, Equipment & Container Appurtenances | Yes/No/NA | Comments |
|---|---|---|
| 12. Do you have the appropriate hydrostatic relief valves installed between every section of liquid piping which can be blocked by manual or automatic valves according to 2-4.7 and 3-2.11 of NFPA 58, 1998 Edition? | | |
| 13. Do you have the appropriate corrosion protection required by 3-2.14 of NFPA 58, 1998 Edition? | | |
| 14. On installations with pumps, are they installed according to 3-2.15.1 of NFPA 58, 1998 Edition?<br><br>On installations with automatic bypass valves, are they installed on the discharge of your pump according to 3.2.15(b)1 and 2-5.2 of NFPA 58, 1998 Edition? | | |
| 15. On installations with compressors, are they installed according to 2-5.3 and 3-2.15.2 of NFPA 58, 1998 Edition?<br><br>On installations with compressors, is there either an integral means of preventing liquid from entering the compressor or a liquid suction protection trap according to 3-2.15.2(b) of NFPA 58, 1998 Edition? | | |
| 16. Do your compressor and pump motors conform with 2-5.1.4 of NFPA 58, 1998 Edition? | | |
| 17. On installations with liquid strainers, are they installed on the suction of your pump or meter according to 3-2.15.3 and 2-5.5 of NFPA 58, 1998 Edition and capable of being cleaned? | | |
| 18. On installations with flexible connections on pumps, compressors or loading and unloading bulkheads, are they installed as specified by 2-4.6 of NFPA 58, 1998 Edition? | | |
| 19. Do you have either excess flow valves, backflow check valves or internal valves as specified by 2-3.3.3 and 3-3.3.7 of NFPA 58, 1998 Edition? | | |
| 20. Do you have container appurtenance protection as specified in 2-3.7 of NFPA 58, 1998 Edition? | | |
| 21. Do you have manual valves and emergency shutoff valves as required by 2-4.5.4, 3-2.10.11, 3-3.3.7 and 3-3.3.8 of NFPA 58, 1998 Edition? | | |

| Piping, Equipment & Container Appurtenances | Yes/No/NA | Comments |
|---|---|---|
| 22. On installations with vaporizing equipment, is it installed according to 2-5.4 and 3-6 of NFPA 58, 1998 Edition? <br><br> Have the liquid traps, temperature controls, and interlocks been tested per the manufacturer's guidelines? | | |
| 23. On installations with regulators, are they installed according to 2-5.7 and 3-2.7 of NFPA 58, 1998 Edition? | | |
| 24. Do you have a breakaway stanchion as required by 3-9.4.2 of NFPA 58, 1998 Edition? | | |
| 25. On installations with swivel-type piping, are they installed as specified by 3-2.10.11(a) of NFPA 58, 1998 Edition? | | |
| 26. Are all above ground lines securely fastened to structural members of adequate strength and supported at proper intervals? | | |
| 27. Are pressure gauges located so that they will not be exposed to physical damage? | | |
| 28. Are there sufficient lines for all purposes, without improper dual use or make-shift connections being used for some operations? | | |
| 29. Are hoses the correct type for each use? | | |
| 30. Are hose couplings of the correct type and properly attached (fully seated on the hose)? | | |
| 31. Is adequate transfer hose storage provided? | | |
| 32. Are the written transfer, loading & unloading instructions available (see § 68.52 of this model program)? | | |
| **Human Factors** | **Yes/No/NA** | **Comments** |
| 1. Have your operators been trained on the written operating instructions for this propane storage facility (see § 68.54 of this model program)? | | |
| 2. For operators on the job on or before June 21, 1999, do they have the required knowledge, skills and ability to perform their duties safely? | | |

| Human Factors | Yes/No/NA | Comments |
|---|---|---|
| 3.  Are your operators whose job duties require the use of the above listed equipment understand the operating limits of the system in regards to:<br><br>Capacity?<br><br>Pressure?<br><br>Temperature?<br><br>Adverse Weather or Natural Conditions? | | |
| 4.  Have your operators been trained in the correct response to conditions which exceed the operating limits of the system? | | |
| 5.  Have your operators been trained in their duties for emergency conditions?<br><br>Fire?<br><br>LP Gas Release?<br><br>Severe Weather or Natural Conditions? | | |
| 6.  Are the written operating instructions available to the operators (see § 68.52 of this model program)? | | |
| 7.  Do the written operating instructions reflect current operation of the facility (see § 68.52 of this model program)? | | |
| 8.  Have major modifications to your propane storage facility taken place (see § 68.48 of this model program)? | | |
| 9.  Are contractors used at the facility? | | |
| 10. Are safe work practices such as lock/tag, hot work and line opening followed at the facility? | | |

| Human Factors | Yes/No/NA | Comments |
|---|---|---|
| 11. Is there a written emergency response plan (see Subpart E of this model program)?<br><br>Is it current?<br><br>Have your operators been trained?<br><br>Do you provide emergency response equipment?<br><br>Has it been checked? | | |

| General Hazards | Yes/No/NA | Comments |
|---|---|---|
| 1. Does your propane storage facility have protection against tampering as specified in under 3-3.6 of NFPA 58, 1998 Edition? | | |
| 2. Does your propane storage facility have lighting as specified in 3-3.7 of NFPA 58, 1998 Edition? | | |
| 3. Is the area around your containers and transfer piping free of all combustible material? | | |
| 4. Has a fire safety analysis been performed for your propane storage facility as suggested by 3-10.2.2 and 3-10.2.3 of NFPA 58, 1998 Edition? | | |
| 5. Has your fire safety analysis been reviewed by your local fire authority? | | |
| 6. Has your facility been required by your local fire authority to provide special protection?<br>Fixed Water Sprays/Monitor Nozzles?<br>Insulating Coatings?<br>Mounding/Burial?<br>Other types? | | |
| 7. Has a federal, state or local agency or fire authority required:<br><br>Local Gas Detection Monitors?<br>Perimeter Gas Monitors and Public Alarms? | | |

| This Hazard Review was Completed by: | On (Date): |
|---|---|
| | |

The latest date by which all changes resulting from the process hazard review are expected to be completed is _____.

Date

## OPERATING PROCEDURES REQUIREMENTS

| Procedure Requirement | NPGA | Certified Employee Training Program (CETP) | NPGA Bulletin | |
|---|---|---|---|---|
| (1) Initial Startup: | Distribution Systems Operations | "Preparing Propane Storage Containers for Installation." | | |
| | Distribution Systems Operations | "Identifying Procedures Used to Pressure Test and Leak Check New Propane Distribution Systems." | | |
| (2) Normal Operations: | Transfer Systems | "Identifying Propane Pumps and their Operation." | | |
| | Transfer Systems | "Identifying Parts and Devices Basic to Compressors." | | |
| | Plant Operations | "Filling Propane Storage Containers." | | |
| | Plant Operations | "Unloading a Propane Transport." | | |
| | Plant Operations | "Unloading a Propane Tank Car." | | |
| | Propane Delivery | "Filling Cargo Tanks on Bulk Delivery Vehicles." | | |
| (3) Temporary Operations: | Plant Operations | "Removing Propane from Stationary ASME Tanks and DOT Cylinders." | | |
| (4) Emergency Shutdown and Operation: | | | 200-89 | How to Control LP-Gas Leaks & Fires |
| | | | 202-81 | What you should do in case of accidents involving LP-Gas |
| | | | 202-93 | Steps to Take in the Event of an Accident Involving Propane |
| | | | 204-88 | How to Handle LP-Gas Fires with Portable Fire Extinguisher |
| | | | 206-91 | Emergency Response Guidelines |

March 4, 1999

| | Procedure Requirement | NPGA | Certified Employee Training Program (CETP) | NPGA Bulletin | |
|---|---|---|---|---|---|
| | | | | 207-94 | Guidelines for Developing Plant Emergency Procedures |
| | | | | 211-91 | LP-Gas Fire Control and HAZ MAT Training Guide |
| (5) | *Normal* (Manual) *Shutdown:* | | | | |
| (6) | *Startup following a normal or emergency shutdown or a major change that requires a hazard review: See item (1) above.* | | | | |
| (7) | *Consequences of deviations and steps required to correct or avoid deviations.* | | | | |
| (8) | *Equipment inspections.* | | | | |

## MAINTENANCE TRAINING

| Module | Certified Employee Training Program (CETP) |
|---|---|
| Basic Principles | Identifying the Proper Use of Personal Protective Equipment |
| Basic Principles | Identifying the Proper Use of Tools and Equipment |
| Basic Principles | Identifying by Sight Hand Tools Commonly Used by Service Technicians |
| Basic Principles | Identifying Pipe/Tube, Pipe/ Fittings, and Associated Tools |
| Transfer Systems Operations | Identifying Propane Pumps and Their Operation |
| Transfer Systems Operations | Identifying the Standards for Sizing, Installing and Inspecting Pump Protective Devices |
| Transfer Systems Operations | Troubleshooting Propane Pumps and Metered Delivery Systems |
| Transfer Systems Operations | Identifying Parts and Devices Basic to Compressors |
| Transfer Systems Operations | Maintaining and Troubleshooting Compressors |
| Distribution Systems | Identifying the Operating Characteristics of Propane Vapor Regulators and Metering Systems |
| Distribution Systems | Installing Propane Vapor Regulating and Metering Systems |
| Distribution Systems | Sizing Pipe for Use in Low Pressure Propane Distribution Systems |
| Distribution Systems | Identifying Steel/Wrought Iron Piping Materials and Installing Procedures |
| Distribution Systems | Identifying Tubing Materials and Installing Procedures |
| Distribution Systems | Identifying and Sizing Vaporizer Systems |
| Distribution Systems | Sizing Propane Liquid Piping Systems |
| Distribution Systems | Identifying Installing and Servicing Procedures for Vaporizer Systems |
| Plant Operations | Identifying the Operating Characteristics of Pressure Relief Valves |
| Plant Operations | Identifying and Installing Gauges in Propane Storage Containers |
| Plant Operations | Identifying the Operating Characteristics of Check Valves |
| Plant Operations | Identifying the Operating Characteristics of Service Valves |
| Plant Operations | Installing Valves in Propane |
| Plant Operations | Inspecting Servicing and Maintaining Container Valves |

| Module | Certified Employee Training Program (CETP) |
|---|---|
| Transfer Systems Operations | Identifying the Operation and Maintenance of Withdrawal Valves |
| Transfer Systems Operation | Identifying the Operation and Maintenance of Bulkheads and Emergency Shutoff Valves (ESVs) |
| Distribution Systems Operations | Installing Propane Liquid Distribution and Vaporizer Systems |
| Transfer Systems Operation | Identifying the Operation and Maintenance of Hoses, Hose End Valves, and Hose Reels |
| Distribution Systems Operations | Identifying the Causes of Corrosion on Metal Surfaces |
| Distribution Systems Operations | Identifying Methods and Procedures Used to Protect Metal Structures from Corrosion |
| Distribution Systems Operations | Identifying Procedures Basic to Installing Anodes and Testing Cathodic Protection Systems |
| Distribution Systems Operations | Identifying Procedures Used to Pressure Test and Leak Check New Propane Distribution Systems |

## MAINTENANCE INSPECTION CHECKLIST AND TESTS FOR PROPANE STORAGE FACILITIES

| I. Construction Code Compliance | | Yes | No and Comment |
|---|---|---|---|
| a) | Check manufacturer's data plate. Is it securely attached and legible? <br><br> For each fixed storage vessel? <br><br> On installations with vaporizers, for each vaporizer? | | |
| b) | Are the data plate(s) free of corrosion? | | |
| **II. Conditions of Container(s) and Vaporizer(s) and Paint** | | **Yes** | **No and Comment** |
| a) | Are above-ground containers properly painted? <br><br> Fixed storage tanks? <br><br> On installations with vaporizers, the vaporizers? | | |
| b) | Are containers and vaporizers free of corrosion damage, dents, gouges, or other damage? | | |
| **III. Foundations** | | **Yes** | **No and Comment** |
| a) | Are foundations in good condition? | | |
| b) | Are footings free of settling which might cause misalignment or piping strain? | | |
| c) | Are containers and vaporizers free of corrosion at masonry contact area? | | |
| d) | Are saddle pads in good condition? | | |
| **IV. Container Connections** | | **Yes** | **No and Comment** |
| a) | Have excess flow and back flow check valves been recently checked for proper operation? | | |
| **V. Tank Fittings** | | **Yes** | **No and Comment** |
| a) | Are all ACME (or other type) connectors in good condition with good gaskets and are they plugged or capped? (See NPGA Bulletin #134 "Care and Inspection of ACME Threaded Hose Couplings.") | | |
| b) | Are all unused openings plugged or capped? | | |
| c) | Are all fittings and hoses leak free? | | |

C-15

| | | Yes | No and Comment |
|---|---|---|---|
| a) | Check manufacturer's data plate. Is it securely attached and legible?<br><br>For each fixed storage vessel?<br><br>On installations with vaporizers, for each vaporizer? | | |
| b) | Are the data plate(s) free of corrosion? | | |
| d) | Are all hoses marked "for LP-Gas service" with a pressure rating of 350 psig (see NPGA Bulletins # 107-91 and #121-89)? | | |
| e) | Are all hoses properly secured, protected, and in serviceable condition and are dust caps on delivery hoses when not in use? | | |
| f) | Are all hoses free from cuts or abrasions that expose the reinforcing fabric and free from soft spots or bulges when under pressure and without kinks, dents or flat spots? | | |
| **VI. Gauges** | | **Yes** | **No and Comment** |
| a) | Are pressure gauges in good condition and are they suitable for 250 psig service (such as 0-400 psig)? | | |
| b) | Are thermometers in good condition and checked for accuracy? | | |
| c) | On installations with vaporizers having temperature controls, are they in good condition and have they been tested in accordance with manufacturer's recommendations? | | |
| d) | Are liquid level gauging devices approved for the service involved and in good condition? | | |
| e) | On installations with vaporizers having level control devices, are they in good condition and have they been tested in accordance with manufacturer's recommendations? | | |
| **VII. Pressure Relief Valves** | | **Yes** | **No and Comment** |
| a) | Is the relief valve data plate legible? | | |
| b) | Do relief valves or vent stacks have protective caps or closures to prevent entry of foreign matter? | | |
| c) | Are weep holes for moisture drainage open and is gas impingement on the container avoided? | | |

March 4, 1999

| | | Yes | No and Comments |
|---|---|---|---|
| a) | Check manufacturer's data plate. Is it securely attached and legible?<br><br>For each fixed storage vessel?<br><br>On installations with vaporizers, for each vaporizer? | | |
| b) | Are the data plate(s) free of corrosion? | | |
| d) | Have the relief valves on containers larger than 2000 gallons and on vaporizers been tested or replaced within the last 10 years as per NFPA 58, 1998 Edition recommendation in E-2.3.2? | | |
| e) | Does external visual inspection of the relief valve discharge indicate no corrosion or obstruction? | | |
| **VIII.** | **Emergency Shut-off Valves** | **Yes** | **No and Comments** |
| a) | Are valves in good condition and do they shutoff tightly? | | |
| b) | Does the emergency shutoff control system function properly? | | |
| c) | Are the remote shutoff controls installed in an accessible area away from the transfer area? | | |
| d) | Are the shutoff controls clearly identified? | | |
| e) | On installations with vaporizers having automatic shutoff controls, are they accessible, identified and been tested according to manufacturer's recommendations? | | |
| f) | Are the emergency shutoff valves and manual transfer valves on your loading or unloading stations protected from pull away damage by a break-away-stanchion. | | |
| **IX.** | **Presence of Combustibles** | **Yes** | **No and Comments** |
| a) | Is the area within 10 ft. of the container(s) and vaporizers free of weeds, long grass, rags, paper, wood or other combustible debris? | | |
| **X.** | **Pipe (for Fixed Storage Tanks and Vaporizers)** | **Yes** | **No and Comments** |
| a) | Are all connections tight? | | |
| b) | Are there sufficient lines for all purposes, without dual use, or are make-shift connections being used for some purposes? | | |
| c) | Are connections labeled "liquid" or "vapor"? | | |

| | | Yes | No and Comments |
|---|---|---|---|
| a) | Check manufacturer's data plate. Is it securely attached and legible?<br><br>For each fixed storage vessel?<br><br>On installations with vaporizers, for each vaporizer? | | |
| b) | Are the data plate(s) free of corrosion? | | |
| d) | Are there visible signs of exterior corrosion? | | |
| **XI. Valves (for Fixed Storage Tanks and Vaporizers)** | | **Yes** | **No and Comments** |
| a) | Are valves in good working order? | | |
| b) | Do seats shut off tightly? | | |
| c) | Is packing free of leaks? | | |
| d) | Are necessary valve handles available at the valve location? | | |
| **XII. Hydrostatic Relief Valves (for Fixed Storage Tanks and Vaporizers)** | | **Yes** | **No and Comments** |
| a) | Are the valves in good working order and not leaking? | | |
| b) | Are the valves fitted with protective caps? | | |
| c) | Are the valve discharges positioned to avoid impinging gas on the tank? | | |
| **XII. Transfer Areas** | | **Yes** | **No and Comments** |
| a) | Are hoses in good condition and free of deterioration, wear, and blisters? See NPGA Bulletin #114 "Guide to Hose Inspection." | | |
| b) | Are hoses capped or plugged when not is use? | | |
| c) | Are hose couplings properly attached and fully seated on the hose? | | |
| d) | Are hose couplings worn or damaged? | | |
| e) | Are coupling gaskets in good condition? | | |
| f) | Are correct coupling wrenches available? | | |
| g) | Are excess flow valves operating correctly? | | |
| h) | Are the loading and unloading risers protected from traffic? | | |
| I) | Are chock blocks provided for rail cars? | | |
| j) | Have your fire extinguishers been tested and/or serviced? | | |

| | | Yes | No and Comments |
|---|---|---|---|
| a) | Check manufacturer's data plate. Is it securely attached and legible?<br><br>For each fixed storage vessel?<br><br>On installations with vaporizers, for each vaporizer? | | |
| b) | Are the data plate(s) free of corrosion? | | |
| k) | Is adequate transfer hose storage available? | | |
| m) | Are the written transfer instructions readily available? | | |
| **XIV. Pumps and Compressors** | | **Yes** | **No and Comments** |
| a) | Are the shafts free of leaks? | | |
| b) | Are pumps equipped with a spring loaded by-pass valve where required? | | |
| c) | Is the by-pass valve functioning properly? | | |
| d) | Are drive belts or couplings protected by suitable guards? | | |
| e) | Is the compressor crank case oil at the proper level? | | |
| **XV. Electrical Equipment** | | **Yes** | **No and Comments** |
| a) | Do all switches, etc. function properly? | | |
| b) | Are all housings properly assembled to maintain seal? | | |

| Inspected By: | Inspection Date: |
|---|---|
| | |
| These procedures were last reviewed or inspected by: | Date: |
| | |

Name (signature)

| Piece of Equipment Inspected: | Most Recent Date: |
|---|---|
| | |

www.ingramcontent.com/pod-product-compliance
Lightning Source LLC
Chambersburg PA
CBHW080645180526
45168CB00008B/3310